WELCOME TO 'THE ULTIMATE CYBER SECURITY GUIDE FOR TIKTOK EVERYDAY USERS, INFLUENCERS AND CONTENT CREATORS'

ISBN:978-1-7636528-1-1

TABLE OF CONTENT

Welcome to 'The Ultimate Cyber Security Guide for TikTok Everyday Users, Influencers and Content Creators' 01

The big issue 06

Speaking the online security language 13

Who threatens your safety on TikTok? 15

How vulnerable is your TikTok account? 24

What are the main online security threats? 33

Protecting yourself with cyber security controls 52

Do TikTok influencers need to care more about online security threats? 69

How different is online security for influencers and content creators? 82

Be aware of regulations 90

The end game 92

About the company 93

Glossary 95

References 100

WELCOME TO 'THE ULTIMATE CYBER SECURITY GUIDE FOR TIKTOK EVERYDAY USERS, INFLUENCERS AND CONTENT CREATORS'

We welcome all TikTokers to the first guide dedicated exclusively to online safety on TikTok. This unique book is packed full of expert information that will make your TikTok experience safer and more enjoyable.

We know that TikTok is a valuable part of your life. We want to make sure it stays that way so you and all the TikTokers around the world can make the most of the opportunity to consume and share content that makes TikTok so iconic.

TikTok is now the premier platform for viewing, sharing, and creating content that's tailored to your hobbies, interests, and passions. It's an exciting place where millions of children, teens, young adults, and adults consume and share content. It's a great place to learn about your local community and issues that are important to you. You can find out about the world around you, catch up with **influencers** and celebrities, and explore the latest cool products.

The amount of content is growing every day as TikTok keeps attracting more users and more creators. There's always something new to see or share. In fact, 34 million new videos are posted every day.

With so much to enjoy, TikTok's popularity just keeps growing. There are now more than 900 million daily users like you. TikTok is the world's fastest-growing social media platform. From 2020 to 2022 TikTok had a **100%** increase in gains compared to Facebook **19.82%**, YouTube **28.1%**. Yet TikTok only started in 2016. Now it has **1.67 billion** members as of May 2024 and users downloaded the TikTok app **4.7 billion** times.

TikTok by numbers

1.67 billion Tiktok members.

900 million daily TikTok users worldwide in 2024.

4.7 billion app downloads.

34 million new videos every day.

Make the most of TikTok - safely

TikTok is a great success story. However, with so many TikTokers now enjoying content around the world, there's a big emphasis on safety for users.

That's because there are people and organisations who want to take advantage of TikTokers. We call them **bad actors** and they could be **hackers**, criminals, or **trolls**. Their aim is not to enjoy TikTok. Instead, they want to cause harm by stealing your personal information, content, or your money, ruining your reputation, or damaging your phone or laptop.

Bad actors like TikTok because they have millions of targets to choose from and they also have many ways to cause harm to TikTokers. If you want to enjoy TikTok safely, you must make sure you're not an easy target. This book will help you to do that.

Welcome to 'The Ultimate Cyber Security Guide for TikTok Everyday Users, Influencers and Content Creators'.

What's this book all about?
And who is it for?

This book is for you. It's written specially for all TikTok everyday users, influencers and **content creators** around the world. We'll show you how you can make the most of TikTok – safely! You'll find the knowledge and guidance you need to protect your TikTok account effectively. Think of it as your personal TikTok safety guide.

You'll get to know the criminals, hackers, and trolls out there. What do they want with you and how do they attack TikTokers?

We'll explain why you need to protect your TikTok account. Throughout the book we'll show you how to stay safe by following the best practices experts recommend.

Staying safe on TikTok – what you need to know

1. Who are the bad actors?

2. What can they do to you?

3. How can you stop them?

4. What's cyber security?

5. What can you do to stay safe online?

This book will help you understand the basic concepts of safety and security on TikTok.

It's not for specialists, it's for TikTokers like you. We want you to have a better understanding of the importance of security. Most of all, we want you to have a better TikTok experience.

THE BIG ISSUE

Why devote a whole book to **cyber security**? We think it's important because there are some serious threats out there. If you don't protect yourself, you could be the next target.

Social media crime is big business. The attacks are increasing every year. Just look at the numbers.

Social media crime is big business

- Cyber crime cost the global economy around **$7 trillion** in 2022. That's expected to rise to **$10.5 trillion** by 2025.

- Cyber attacks increased by over **125%** in 2021 compared to 2020.

- Increasing volumes of cyber attacks continued to threaten businesses and individuals in 2022.

- There are over **2,200** attacks each day. That's nearly 1 cyber attack every 39 seconds or over **800,000** attacks each year.

How about those familiar apps you use every day. How safe are they? Look what happened to users of PlayStation, Yahoo and Google Docs.

Back in 2011, the PlayStation Network was hacked. The attackers accessed personal details of 77 million users. It was the largest security breach of its kind to hit gamers. Gamers faced the risk of having their personal details released. Sony had to shut down PlayStation for over three weeks.

In 2014, a state-sponsored hacker accessed some 500 million Yahoo accounts. They stole everything from names and email addresses to telephone numbers, passwords, and dates of birth.

In 2015, hackers accessed the Ashley Maddison online dating site and copied users' names and addresses. Many of the users were already married and that could prove embarrassing. The hackers threatened to release this personal information unless the owners shut down the site.

In 2024, Google issued a warning against an attack on Gmail accounts. Hackers invited users to view a Google Doc. Clicking the document link gave hackers access to email, contacts, and other documents. The hackers could use that information to scam people or steal money.

In 2024, hackers attacked one of the world's largest mobile and internet services providers. They stole personal data belonging to 73 million current or former customers. They published full names, email addresses, social security numbers, dates of birth, and passcodes on the dark web. When the information is on the dark web, hackers can sell it or use it to plan other attacks.

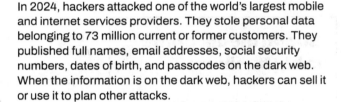

Sometimes, it's personal

Those are attacks on big organisations. You probably think it could never happen to you. The trouble is, it does hit individuals. So, what's the impact on TikTokers like you?

See what happened to these three TikTokers. We've changed their names, but the stories are real. They're familiar situations and they could affect any TikToker.

Mike saw a great offer in an email that claimed to be from TikTok. He clicked the link and enjoyed his reward. Two months later, he found his bank account was 500 dollars lighter. The criminals had used the link to gain access to his phone and steal bank details.

Once criminals had the details, it was easy to take money from his account. In an attack like that, criminals trap victims like Mike first, then wait to do personal damage.

This is a real story about an individual. We've changed the name and some of the details to protect their identity.

Jason is a successful content creator with thousands of followers. He posts new videos every week on TikTok and earns money from sponsors. One of his contacts told Jason he had seen similar videos on other channels. Jason found that his content was appearing on three different channels as well as his own.

He was angry that other people had stolen his content, but he also realised that he was losing revenue. Viewers were watching the videos on other channels, so his number of followers was not increasing. As Jason's sponsorship income was linked to numbers of followers, his income fell. Jason quickly realised the value of protecting his content.

This is a real story about an individual. We've changed the name and some of the details to protect their identity.

Sheila is a TikToker and enthusiastic online shopper. She used a single password for TikTok and all her store accounts. She was shocked when she received a payment demand for 300 dollars' worth of clothes she hadn't ordered. Hackers had stolen the password and used it to access her store account.

Sheila didn't just lose money, she also lost confidence in her ability to shop online safely. After that bad experience, she made sure that she used a different password for every account.

This is a real story about an individual. We've changed the name and some of the details to protect their identity.

Time to close the door

Unfortunately, the experience of Mike, Sheila and Jason is not unusual. It happens to ordinary TikTokers every day and it hits them hard. That's why we want to help you enjoy TikTok to the full without worrying about other people's actions.

Safety starts with stronger **online security.** Security on TikTok is like safety on any other online activity. You're probably cautious about suspicious Internet sites, online banking risks, gaming scams or email fraud. It's just as important to stay safe on your TikTok account.

As our three TikTokers found out, the threats out there are real – and they are damaging. The risks are high, so you must protect your TikTok account. Online security is essential for all TikTokers.

That's why this book is so important. There are millions of articles and videos about online security, but they don't give detailed information on security for TikTok everyday users, influencers and content creators. This book is unique because it's just for TikTokers. Before we get into the detail of online security, let's get familiar with some of the terms you'll come across.

SPEAKING THE ONLINE SECURITY LANGUAGE

Throughout this book, we'll use standard terms that security professionals use. So, here's a brief definition of the most commonly used terms:

Vulnerabilities

Vulnerabilities are weak spots on your phone, laptop, or tablet. Hackers and criminals exploit them to access your device. For example, a weak password - or worse, no password – makes it easy for a hacker to gain access to your account.

Threats

If you've got vulnerabilities, you face cyber threats. You could be exposed to viruses or hit by cyber criminals or trolls. They'll take advantage of a vulnerability to access your devices, personal data, and content.

Risks

So, how big is the likely problem? When we talk about risk, we're saying, how likely is the chance of a threat getting through. By taking steps to protect your account, you lower the risk.

Incidents

An incident is a sequence of events that happen when someone attacks your account. Suppose you get an innocent looking message with a link to an offer. You click on the link. All seems okay until you realise money is missing or your personal data appears on a rogue website.

Exposure

Exposure means you're liable to be affected by a security incident because of those vulnerabilities.

Impact

Just how bad is the likely damage? That's what we call impact – the level of damage you might suffer. It might just be loss of personal information. An attack could lead to blackmail, ransom demands, damage to your reputation, and more.

Cyber security controls

Cyber security controls are a combination of technologies, processes, and best practices. They act like defense barriers to protect you. We'll go through each of the controls in detail later.

Those are the terms we'll use throughout the book. The first term we'll talk about is threats and the people or organisations who pose them.

WHO THREATENS YOUR SAFETY ON TIKTOK?

There are three types of people who might threaten your TikTok account –
serious criminals,social media hackers, and trolls.

Serious criminals

Criminals are sometimes known as 'bad actors'. We're talking about people who intend to do harm. They're likely to be individuals or gangs stealing money. They use the money to fund drugs, child sexual exploitation, terrorism, or other criminal activities. They mainly attack businesses where they can 'earn' more money from blackmail, ransom demands, or stolen financial details.

There are also **'state actors'**. They operate with the backing of their governments. Their targets are individuals, businesses, or governments. They might want to discredit someone, gather large amounts of personal information, or maybe just steal money. Sometimes, they attack for political reasons – interfering in elections or try to harm politicians or governments.

From Russia without love

In April 2024, the US Cybersecurity and Infrastructure Security Agency (CISA) identified Russian government-backed hackers.

They used access to Microsoft's email system to steal correspondence between government officials and Microsoft. They also used the emails to try to break into other Government agencies' systems.

Those are the criminals who make the headlines. Fortunately, big-time criminals are less interested in TikTok everyday users like you. They stick to attacks on bigger targets. More often we find that everyday attacks come from small-time criminals, trolls, or social media hackers.

Social media hackers

Hackers find social media accounts like TikTok a very attractive target to attack. According to the website Station X, social media hacking is now commonplace. The website published some worrying statistics.

Hackers use attacks to steal money or access personal information. Some hackers blackmail their victims or demand a ransom to free their accounts.

Other hackers trick victims into sending money or use stolen accounts to **scam** other people. They also use stolen information to create fake identities or sell data to other criminals.

That's why selling a stolen account is profitable. PC Magazine reported that the price for a stolen TikTok account is quoted as $8. That's cheaper than an Instagram account at $12 or a Facebook account at $14. It's a real bargain compared to a LinkedIn account that sells for $45 on the dark web.

Not all hackers are after money. Some access other people's social media accounts 'just for fun' to demonstrate their skills.

Hackers love social media

- On average **1.4 billion** social media accounts are hacked every month. That number will continue to rise as more and more people create social media accounts.

- From 2021-2022 the number of social media accounts hijacked increased by **1,000%.**

- A Google report found that **20 %** of social media accounts will be compromised at some point.

Want to buy a stolen social media account?

🔴 Reddit $6

⚫ Tiktok $8

📷 Instagram $12

🔵 Facebook $14

🔵 LinkedIn $45

Trolls and cyber bullying

- It's a sad fact that **44%** of Australian young people reported having a negative online experience in the last 6 months of 2023.

- This includes **15%** who received threats or abuse.

- Roughly **40%** of Americans have experienced some form of cyber bullying in the form of harmful or damaging comments, posts, or messages.

Trolls posting harmful or negative messages can cause personal distress. The website CORQ published comments by several TikTok users. Many of them claimed that trolling on TikTok was harsher than other social media. Trolls posted negative comments about TikTokers' looks, their clothes, and their content.

Trolling and cyber bullying take many forms, from hate-filled comments and derogatory messages to personal attacks and character assassination. These actions are driven by a desire to undermine, humiliate, and provoke a reaction from victims.

Some users just dismiss the trolling, but it can be more damaging. Unfortunately, there are too many stories of worry, mental illness, even suicide from trolling. They happen to ordinary people like 'Penny'.

Penny was an outgoing teenager with thousands of followers on TikTok who loved her daring fashion sense. She had a shock when an anonymous troller sent messages containing fake compromising photos featuring her face. The troller also shared her address and sent threatening messages. Penny suffered high stress levels and for months was afraid to go out on her own. Although she started to regain confidence, she worried that the trolling could start over again.

This is a real story about an individual. We've changed the name and some of the details to protect their identity.

Is TikTok safe?

Many commentators are concerned about TikTok's data privacy practices. TikTok, like other social media companies, collects data on users. The worry is how they store and share the data.

Here's a brief outline of what happens to you and your data when you use TikTok:

- TikTok collects data on your search and browsing history, demographic information, your likes, and the people you follow. They also capture your facial identity, voice prints, texts, location, and photographs. This helps TikTok build a virtual identity of you and your interests.

- TikTok shares user data with advertisers. The advertisers analyse the data to understand how you use TikTok. They learn where you are and what your interests are. They then use the information to create ads that they think you will be interested in. That's known as targeted advertising.

- TikTok shares data with other social media businesses to learn more about user activities on other apps.

- TikTok shares user data with other third-party apps.

- Law enforcement agencies and federal agencies can request user data from TikTok. They use the data to assist with investigations.

TikTok takes the same safety and security measures as all social media companies. However – and it's a big BUT – many individuals

and governments worry about TikTok's relationship with the Chinese government. Does TikTok share data with the government?

The US government were concerned about the risk of spying on individuals and government agencies. In spring 2024, legislation was passed by more than eighty percent of representatives in Congress and signed by President Biden. The app will be removed from distribution in U.S. app stores unless ByteDance sells it to another entity. Currently, ByteDance is appealing against the ban as of May 2024.

TikTok responded to the bill and set out its position on its relations with the Chinese government. ByteDance stated that they do not have a relationship with the Chinese government. They pointed out that:

- TikTok executives are based in local markets, not in China.

- TikTok has a diverse board with members from different countries.

- TikTok has a range of measures in place to protect users' data.

- TikTok stores users' data in different territories.

Avoid threats to your safety

Do you really want people stealing your personal information and going on to rob your bank account? Do you want people trashing your reputation, changing your content, or sending fake messages to your contacts?

There are simple things you should do like changing passwords, adding a phone number, or switching off comments. We'll explain those and many more security measures to protect you later. Before that, we'll tell you more about the possible vulnerabilities that open the door to attacks or trolling.

HOW VULNERABLE IS YOUR TIKTOK ACCOUNT?

Remember our definition of 'vulnerabilities'. They're the weak spots on your phone, laptop, or tablet. Bad actors exploit them to access your device. Just one example - you've got a weak password, or worse, no password. That opens the door to hackers.

Think of a vulnerability as the equivalent of a smoke alarm with a flat battery, a door left unlocked, or a burglar alarm switched off. You get the idea. So, what are the possible vulnerabilities when you're online?

Password weakness

Weak passwords are a big risk

- **80%** of successful data breaches are the result of weak passwords.

- **30%** of people still use simple passwords to protect their accounts.

- **34%** of people repeat variations of the same password.

Password weakness is one of the most common vulnerabilities. Hackers try to steal passwords by encouraging victims to click on fake links in TikTok messages or send personal information. They also attack TikTokers' devices directly by inserting harmful codes or software that allow them to gather information. One form of software captures users' **keystrokes**, which can reveal passwords or other confidential information.

How strong are your passwords and how often do you change them? Sometimes, online operators specify what a password should include – total length, so many numbers, so many symbols, and so many capital letters. When you create your password, you generally get a message like strong, weak, or try again.

Part of the problem with passwords is that you should be able to remember them. So, if a company gives you an initial password like **4zT*2x3$9M#,** that's unlikely to stick in your memory. You get the option to change the password, so what do you choose?

Choose a strong password

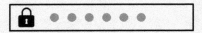

The best advice is to make the password as complicated as possible. It could be very long, or it could include many different combinations of random numbers and letters. Hackers prefer a password that's short and simple because they can work it out quickly. If it takes too long, hackers will move onto another target.

One way to make passwords strong is to use an online **password generator**. They produce some very complex passwords that hackers will hate. Because they're complex and hard to remember, you'll also need a **password manager** like **NordPass**[1] to use them.

A password manager is an app on your phone or computer that stores your passwords. You login with a master password to enter and choose an account. The password manager will remind you of the right password.

That combination of numbers, letters, and symbols is a great starting point, but there are some popular combinations you should avoid.

Passwords to avoid ○○○

- Your birthday – it's one of the first terms hackers try. Your birthday appears on your social media and your sign-up data, so it's easy to find.

- Your name – no prizes for guessing why that's easy to work out.

- Your pet's name – especially if the pet regularly makes guest appearances on your account.

- 123456 or similar number combinations. It's ranked as number 1 password to avoid!

- Admin – that also ranks high on the popularity charts.

- ****** - it's meaningless, but worth a try for hackers.

[1] https://go.nordpass.io/aff_c?offer_id=488&aff_id=104853&url_id=9356

Don't reuse passwords

Once you've created a strong password, it's tempting to use the same one on all your accounts. Don't go there! If a hacker works out the common password, you've got more than one problem.

Make sure your TikTok password is different from all the others.

Don't share passwords

Don't give your passwords to anyone else – even if it's someone you know and trust. They might not protect the password the way you do, and that good gesture will come back to haunt you.

Change passwords

You might have a great password, but don't plan to use it for life. Hackers keep trying to attack so the longer you leave it, the bigger the risk. Plan to change your passwords every three months at a minimum.

Passwords does and don'ts

- Create strong passwords
- Use a password generator and manager
- Avoid common passwords
- Don't reuse passwords
- Don't share passwords
- Change passwords regularly

Email

Emails can also be a source of trouble. Hackers can work out some personal information from your email account. They use that information to plan an attack on your TikTok account.

Keep personal data out of emails

91%

of all cyber attacks begin with a scam email.

You answer an innocent-looking email that claims to be from TikTok. It asks you to reset a password or provide some personal information. The email might be about a competition, a survey, or a special offer.

Be cautious! The email is probably a scam to gather information for future attacks on your TikTok account. You might not have an immediate problem, but scammers could just be biding their time.

Check carefully. Is the TikTok address, correct? It's easy to check. Does the logo look wrong? There are always clues to help you identify a scam email.

Operating system problems

Hackers are good at finding weaknesses in the **operating systems** that run your phones, computers, or other devices. If they find a weakness, they will use it to access your system, cause damage, or steal data.

You can't do much about an operating system vulnerability. That's the responsibility of the manufacturer. When manufacturers find a weakness, they try to fix it by applying 'patches' or system updates.

Some systems include automatic updates. In some cases, manufacturers notify users and ask them to apply the latest updates. These updates are essential. If you don't apply them, you are at risk from new threats.

Wi-Fi security

We've spoken about the measures you should take to avoid risks. Sometimes, you can be vulnerable to attacks that come via Wi-Fi. If you're away from home, you connect at a public Wi-Fi hotspot.

Be cautious, there's no guarantee the connection will be secure. If it isn't secure, hackers could be listening in ready to intercept any useful information.

If you're connecting at home, how secure is your Wi-Fi system? A lot of people believe that the Wi-Fi service provider secures the connection. Not necessarily. So, it pays to take a few simple steps to protect your account and all the devices connected to the Wi-Fi network.

For a start, change the name of your connection. Manufacturers give the Wi-Fi router a default name. Change that, following the set-up instructions for your router. Then create a strong password for the network – and don't forget to change it regularly.

Use Wi-Fi carefully

Be cautious at public hotspots.

Change the name of your home Wi-Fi.

Create a strong password.

Human error

It sounds unlikely, but security professionals recognise that human error is one of the biggest causes of security problems. A study by Stanford University and security firm Tessian has found that 88 percent of data loss incidents are caused by employee mistakes.

That error could be as simple as accidentally pressing the wrong key. You release personal information that you didn't want to disclose. Sometimes, the mistakes are caused by lack of care or neglect. For example, you click on a scam link in a message, forget to update your device with the latest security information, or give away your password without thinking about the consequences. Those simple errors give hackers the information they need to attack your account.

88%

of all data loss incidents are caused by human error.

Avoid the threats

These are vulnerabilities that criminals and hackers are aware of. That's why it's essential to take the right cyber security measures. Before we describe them, let's look at some of the threats you could face if you ignore vulnerabilities.

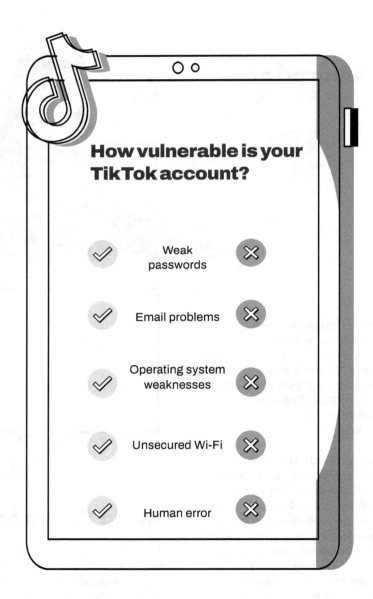

WHAT ARE THE MAIN ONLINE SECURITY THREATS?

Now you understand vulnerabilities, you can see why it's important to take online security seriously. Vulnerabilities expose you to damaging cyber threats. There are many ways this can happen.

All about me –
don't lose your personal information

Personal information is gold dust to criminals and hackers. If your account is vulnerable, they will gain access to your personal information. Then they use the stolen data to do some serious damage:

- **Impersonate** your identity and scam your contacts.

- Steal money.

- Pass personal information to other parties.

- Release confidential or embarrassing information.

The cost of identity theft

- The estimated cost of identity crime in Australia is over **$3 billion.**

- An estimated **15 million** Americans had their identity stolen in 2021.

- Identity thieves stole around **$52 billion** from Americans in 2021.

Tyler's story highlights the cost of losing your personal information.

Tyler signed up for an amazing offer that came through a TikTok message. It seemed too good to be true, but what's the risk, he thought. All he had to do was confirm a few personal details like his username, age, and favourite things to do as well as providing an up-to-date profile photo.

Unfortunately, the offer was a scam. A few weeks later, he was still waiting for the offer to turn up. Instead, he got some angry messages from contacts who had been scammed by someone using those personal details to impersonate him.

This is a real story about an individual. We've changed the name and some of the details to protect their identity.

Phishing trips

Phishing is the term we use to describe attacks to gather information. Attackers send emails or messages that contain links to scam websites. The websites contain malware or viruses that will damage your devices. The website might also try to trick you into revealing personal information or transferring money.

If hackers get hold of your information, your online life could turn into a scary roller-coaster ride. Sometimes, the results are not so dangerous - just downright annoying.

Phishing is growing fast

- A new phishing website is created once every 20 seconds on average.

- Phishing attacks account for more than **80%** of reported security incidents.

How Jodie fell victim to a phishing scam.

Jodie saw an online survey that looked appealing – 'How cool are you?' The survey asked lots of questions about her age, music tastes, favourite clothes stores, and influencers she followed. Jodie thought it was a great chance to impress!

What followed was not so great. For weeks after she completed the survey her screen was bombarded with pop-up ads. Her inbox was flooded with special offers – all from different companies. Okay, it wasn't damaging but it proved difficult to turn off.

That survey had given hackers valuable personal and demographic information that they could sell to many different advertisers. The advertisers didn't realise how many companies were using the same stolen information. They were unaware of the impact on Jodie.

This is a real story about an individual. We've changed the name and some of the details to protect their identity.

Getting to know you – be careful about sharing personal information

Social engineering could hurt you

- **98%** of cyber attacks involve some form of social engineering.

- **41%** of higher education cyber security incidents started with social engineering.

Criminals don't always use technical vulnerabilities to find out your personal information. Criminals and hackers collect information by persuading you to trust them and take actions that will ultimately harm you. This process is known as **social engineering**.

For example, a criminal pretends to be from an organisation or person you trust. If the deception works, the attacker will encourage you to take further action or give away sensitive information like passwords, date of birth, or bank details. Attackers might also encourage you to visit a website where they have installed **malware** that can damage your devices or give them access to information on your devices.

Attackers can also gather information on you by monitoring your post and comments on TikTok. They use the information to build a profile they can use to create damaging attacks.

The message is, be careful what you share.

Mike's story illustrates the risks of sharing too much personal information on TikTok.

Mike loves the outdoor life, and he has a favourite username to match - @mikethehike. Mike shared images of hikes through national parks. He created TikToks wearing his favourite climbing gear. He posted dates of his bike racing plans for the year and shared posts from his friends about their favourite adventures. Mike was surprised when several of his friends called to say they had been infected by a virus.

Hackers had found the username @mikethehike on some of his posts. They used that information to work out various possible password combinations and eventually broke into his laptop. Then they sent messages containing viruses to all his contacts. Although the viruses did not cause any serious harm, Mike worried that he had betrayed the trust of his friends.

This is a real story about an individual. We've changed the name and some of the details to protect their identity.

Trolls in the wings

Your TikTok profile and your posts can come back to bite you. The more you share, the more people know about you. Maybe they like what you do, and they let you know. Unfortunately, the comments aren't always what you want to hear. They could be insults, comments about your appearance, or your lifestyle.

Some trolls use TikTok to bully you by making you feel inferior or subjecting you to threats. You receive a constant flow of negative, damaging messages that make your life miserable day after day.

Trolling and cyber bullying are extremely distressing. They affect your mental health, causing anxiety, depression, and loss of self-confidence. In the worst cases, it has led to suicides.

Trolling is a threat you need to stop quickly before it gets out of control. That's not always easy. Some trolls are anonymous, so it's hard to respond or control the comments. However, that's not always the case. The trolls could be people you know from school or work. They might be people you thought were friends, but now are making your life stressful.

Trolling is hard to avoid

- **38%** of online trolls target people on social media.
- **46%** of Internet users between the ages of **13** to **17** years have faced online harassment.

'Tracey' worked hard to get more likes. Then negative comments ruined her life.

Tracey loves fashion. She dresses up in some of her favourite clothes for TikToks and gets plenty of likes from her friends. When the comments started getting nasty, it wasn't so great. Trollers posted comments on her clothes, her weight, and her make-up that were very upsetting.

When someone used AI to give her an older appearance, that was enough for Tracey. She turned off comments. It made life more bearable, but she missed the comments from her friends.

This is a real story about an individual. We've changed the name and some of the details to protect their identity.

Infectious moments – avoid virus attacks

Malware infections are growing

- Malware infections have increased by **87%** in the last decade.

- **94%** of malware is delivered by email.

- **5.5 billion** malware attacks were detected around the world in 2022.

If hackers can break into your system, they will damage your devices by infecting them with viruses or malware. Malware is the term used to describe malicious or harmful software.

Malware and **viruses** damage your phone or laptop, so they don't work correctly. The software is also used to steal information from your TikTok account. Hackers might demand money to remove the virus, or you might find that personal information has been stolen without you realising.

'Carrie' gets hit by malware.

Carrie thought her new router was great. It gave her faster access to TikTok, better streaming and amazing gaming on her laptop - and all for a lower price. Unfortunately, she didn't take the security precautions that are essential with a new router - changing the router name and creating a new password to access the Wi-Fi network.

Just a few days in, things started to go wrong. The laptop screen froze, videos were jerky and annoying pop-ups appeared. Even worse, the same problems spread to her brother's laptop and the family computer. A virus had entered the router from the Internet and spread through the household Wi-Fi network.

The family were not happy. Fortunately, Carrie's uncle was a computer tech. He cured the problems on the family devices and changed the router security settings. He gave Carrie a clear message – pay attention to security.

This is a real story about an individual. We've changed the name and some of the details to protect their identity.

Your money or your digital life – ransomware

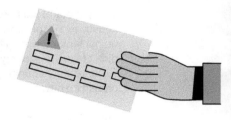

You hear about in the news - criminals hit big business with ransom demands for millions. They install **ransomware** to do that. Ransomware is a type of malware that prevents you from accessing your system. The software could lock your screen or your files until you pay a ransom.

You think it couldn't happen to people like you, but it does. Criminals are happy to earn small change from innocent individuals. If you don't pay up, they block your account, threaten to release personal information about you, or worse.

Ransomware costs big money

- **5073** ransomware attacks in 2023.

- Ransomware payments topped **$1 billion** in 2023.

'Mark' gets hit by ransomware.

Mark was a teen TikToker who liked to visit different websites to get new ideas for creating content. Unfortunately, he strayed into an adult website by accidentally hitting the wrong key. He closed the site quickly enough, but the story didn't end there.

Next day, an anonymous email came through via his TikTok account. "You've got three days to pay $200. If you don't pay, we'll release your browsing habits to your family and friends." An offer like that is hard to refuse!

This is a real story about an individual. We've changed the name and some of the details to protect their identity.

Access denied – keep login details safe

Protect your password

- **44 million** accounts were vulnerable to account takeover due to compromised or stolen passwords.

- Compromised passwords are responsible for **81%** of hacking-related breaches.

Hackers have many ways to gain access to your account. It's known as **compromised login**. The attacker obtains your login details or finds a way to bypass your account's security measures. They use phishing attacks, malware infections, weak passwords, or other vulnerabilities to obtain the login details.

Once hackers have access, they can break into your system and take control. They might decide to block access to your account or gather personal information. You may find yourself facing ransom demands for this too.

'Jacqui' gets locked out of TikTok.

Jacqui got home from class and logged into her TikTok account. She got a surprise 'access denied' message on the screen. She tried again three times. No change, then she thought about changing her password. That didn't work either.

Later that day, an email message came through, "Unlucky Jacqui, no way back into TikTok unless you pay us $150. Have a nice day." The hackers had infected Jacqui's laptop with software that prevented access. Jacqui was very cautious after that incident and installed security software to prevent that kind of attack in the future.

This is a real story about an individual. We've changed the name and some of the details to protect their identity.

Clouds on the horizon – check security levels when you use the cloud

Do you keep images and other information in the cloud? Are you confident the operators are protecting your data? Storing material on your own device means you're in control of security. In the cloud, it's out of your hands. So, it pays to ask about cyber security before committing your life's digital memories to the cloud.

There are three main ways criminals can steal your information from the cloud. They try to break into the cloud site. That can be difficult because cloud providers put very strong security measures in place. They might try to break in by stealing your password and impersonating you. Hackers also intercept information when you upload it to the cloud over the Internet.

Cloud security can be a problem

- More than a third of businesses experienced a data breach in the cloud in 2023.

- Human error was reported as one of the leading causes of cloud data breaches.

'Don' put too much trust in cloud security.

Don ran out of storage space on his laptop. He didn't want to waste time cleaning up old TikTok messages and videos – life's too busy. An ad for low-cost cloud storage looked attractive. He signed up and uploaded the content via the Internet.

Cloud providers recommend encrypting content before sending it over the internet. Don was too busy to do that. Hackers intercepted the uploads on the Internet and were able to steal valuable personal information.

This is a real story about an individual. We've changed the name and some of the details to protect their identity.

Lost and found –
be careful with your devices

What happens if you accidentally lose your phone or forget to pick up your laptop if you've used it away from home? You get it back if you're lucky, but you might not get back all the TikTok data and personal information that was on the device.

Unless a criminal has stolen the device, the data will likely be there with nothing changed. However, if the data on the device isn't secured, you could be in for a nasty surprise. Criminals could steal your data or demand a ransom for returning it.

If someone steals your device or your data, let TikTok knows so that they can block your account before criminals use the information to cause other problems.

Don't lose your phone

- **70 million** smartphones are lost or stolen each year.

- Only **7%** are recovered.

- **50%** of phone theft victims would be likely to pay $500 to retrieve their stolen phone's data.

'Erin' lost her phone and her data.

Erin took the subway home from a great show. It was her favourite artist's final farewell tour. The subway was crowded with other people from the same event. Erin had photographed most of the show on her camera. She couldn't wait to share the images on TikTok.

When she got home, panic, no phone! It must be on the subway, surely. Good news, she called the subway depot and asked if they had a phone. She was able to identify it and collect it the next day. The phone was working, but there were no images. Someone had worked out her
password, accessed the phone, and transferred all the photos to their own device.

This is a real story about an individual. We've changed the name and some of the details to protect their identity.

Don't take the risk

Those are just some of the threats you might face if you don't protect your TikTok account. So how do you protect your accounts and your data?

PROTECTING YOURSELF WITH CYBER SECURITY CONTROLS

Cyber security is a combination of technologies, processes, and best practices. You should use them to make your TikTok account more secure.

This section describes TikTok's measures to protect users. It provides some simple tips you should follow to keep safe online and explains the main types of cyber security controls.

How TikTok protects you

This is a summary of the safety and security measures TikTok takes to protect users.

TikTok is committed to making sure that any personal information shared intentionally or accidentally on TikTok does not lead to harm.

TikTok does not allow content that includes personal information that may create a risk of stalking, violence, phishing, fraud, identity theft, or financial exploitation. This includes content that someone has posted themselves or that they consented to being shared by others.

TikTok does not allow posts that include:

- Personal phone numbers and home addresses.

- Financial and payment information, such as bank account and credit card numbers.

- Login information, such as usernames and passwords.

- Identity documentation, cards, or numbers, such as passports, government-issued identifications, and social security numbers.

- Threats or encouragement to share personal information or to hack someone's account.

TikTok does not allow:

- Access to any part of TikTok through unauthorised methods.

- Attempts to obtain sensitive, confidential, commercial, or personal information.

- Any abuse of the security, integrity, or reliability of our platform.

- Sharing malicious files, content, and messages that contain viruses or other harmful materials endangering cyber security.

- Attempts to obtain personal information or access content, accounts, systems, or data through the use of any deceptive technique.

Simple security tips

Security doesn't have to be complicated. Here are some simple tips you should follow to stay safe every time you go online.

Stay safe online

- Don't share personal information on your posts.

- Be careful what you share.

- Set your profile to private rather than public.

- Control who can comment on your posts.

- Control who can send you messages.

- Set up alternative contact methods to recover your login details.

- Store important information in other locations.

- Choose strong passwords.

- Use alternative logins like fingerprint or facial recognition.

- Use 2-step verification to login.

- Login with a passkey.

- Set up trusted lists and blacklists for incoming emails.

- Don't click on suspicious links in emails or messages.

Don't share personal information on your posts

That's the number one TikTok tip. The TikTok guidelines say it and we agree. Personal information is the top target for hackers and criminals. It's valuable to them and it's harmful to you if you lose it.

Hackers could use the information to access your bank account. They could try to scam your contacts, or troll you. Keep your personal information private – don't show phone numbers, passwords, usernames, bank details or email addresses anywhere on your TikTok account.

Be careful what you share

It's tempting to share all the great things in your life on TikTok - keep your friends up to date and show you're a cool person. It's better to take a step back and be cautious. It's not just friends who are watching. We've told you about all the hackers and criminals who could be watching. Even if they don't attack your account, they could troll you.

Set your profile to private rather than public

If you are a user and your **profile is public**, your name, username and profile photo will be visible to anyone on TikTok. That means other people will be able to search for your account. Make your profile private if it's not set up that way.

However, influencers and content creators may need to keep their accounts public, due to the nature of their activities and better engagement with their audience on the platform.

When your profile is public, any TikTok user can view your videos. They can post comments or engage with the content you've created and shared.

If you are a user and you switch to a **private profile**, you can approve or deny requests from people who want to follow you. Only people you've approved as followers will see your content.

If you're a user under sixteen, TikTok makes your profile private automatically. If you're older, your profile will be public. However, you can switch it to private. Only provide the minimum account information when you set up your profile. Usually, that means name, date of birth, gender, and contact information.

That way you limit what people learn about you. As we keep emphasising, hackers can build up a profile from small amounts of information and then they're ready to attack. A public profile gives them a great start.

To set your account to private

Go to your profile page.

> Tap three dots in the right-hand corner and select 'Privacy and Settings'.

> In the 'Privacy and Safety' option, toggle 'Private Account' on/off.

Control who can comment on your posts

Everybody wants to get more likes. It's only natural. When somebody makes a positive comment on your post, you feel even better. What do you do if things turn bad and the trolls hit your account?

Make sure you report any malicious posts. TikTok should then remove them. If you are a user, one easy way to stop it happening again is to set your account to private. Then only let friends or followers you really trust post comments.

However, influencers and content creators should leave comments enabled so that anyone can engage with them.

Control who comments on your TikTok posts

> Go to your profile page.

> Within 'Privacy and Safety', select 'Who can send me comments'.

> Choose 'Friends' to limit comments to people you know.

> You can also turn off comments on individual videos by going to the menu button on the video and selecting 'Comments off'.

Control the messages you want to accept

Trolls and cyber bullies don't just post comments. They send messages too and they can be distressing or even threatening. Even simple malicious messages are upsetting. So, use your privacy settings to limit the friends and followers who send you messages.

Control who sends you direct messages

> Go to your profile page.

> Within 'Privacy and Safety', select 'Who can send messages to me'.

> Choose from the various options – Everyone, Friends or Off.

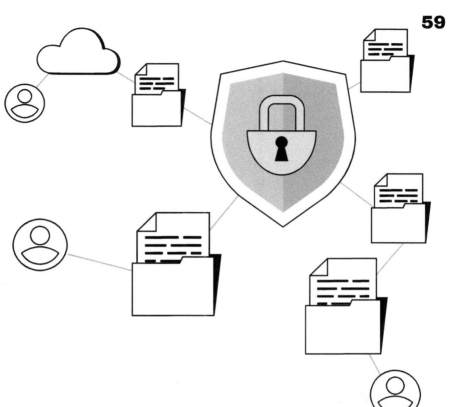

Set up alternative contact methods to recover your login details

What happens if you forget your login details? Do you keep trying to login and eventually get shut out? How do you get your login details back?

Instead, provide TikTok with an alternative contact number or email via the TikTok settings. TikTok can use this to securely send you instructions to regain access. Afterward, it's a good idea to change your login details immediately to ensure your account remains secure.

Store important information in other locations

To protect your personal, private, or confidential information, store it outside of your TikTok account whenever possible.

By keeping sensitive data in secure locations, such as encrypted cloud storage you reduce the risk of exposure in case your account is ever compromised.

Choose strong passwords

A strong password is one of the most important ways to keep your TikTok account secure. You'll find all the password advice you need earlier in the guide. Make life difficult for hackers and they won't waste time on you.

Use alternative logins like fingerprint or facial recognition

Most phones will let you login this way. So why not protect your TikTok account with **fingerprint** or **facial recognition**. Your security levels will rocket. **Biometric credentials**, as they're known, are the strongest form of cyber security. They can't be copied unless hackers have invested in multi-million-dollar systems.

Use 2-step verification to login

Two-step verification means what it says. After you enter a password, you must enter another piece of information. This might be information you hold, or it might be a code that TikTok sends you. These codes are known as One-Time Passwords or OTPs because you can only use them once.

Using 2-step verification makes life more difficult for hackers. They migh be able to work out your password. Trying to guess an OTP is something they won't waste time on.

Turn on 2-step verification

- Check that you have the latest version of TikTok downloaded on your device.

- Tap Profile at the bottom of the screen.

- Tap the Menu button at the top.

- Tap Settings and privacy.

- Tap Security.

- Tap 2-step verification and choose at least two verification methods: email or text.

Login with a passkey

Passkeys are another secure method of logging into your account. A passkey is a key stored on your mobile device. You can use it to login without having to remember a password. Currently, passkeys are only available on Apple devices.

Set up a passkey

- Tap 'Profile' at the bottom.

- Tap the Menu button at the top.

- Tap 'Settings and privacy'.

- Tap 'Account'.

- Tap 'Passkey', then tap 'Set up' on the next screen.

- Follow the instructions to complete setup.

Set up trusted lists and 'blacklists' for incoming emails

When hackers are looking for personal information to attack your TikTok account, they try many different sources. Your TikTok profile is a good starting point, but sometimes your main email account provides additional information.

To control the emails you receive, set up trusted lists of contacts and 'blacklists' of names that you recognise as potential trouble. Suppose an email from a name not on your trusted list arrives. Your system will flag it up as 'junk'. You can decide to delete it, add it to your trusted list or put it on the 'blacklist'. If the name on the 'blacklist' keeps trying, the request will get rejected automatically.

If you do receive an email or message that's not on your trusted list, check it carefully even if it looks genuine. Don't click on any links if they look suspicious.

Remember the tips

These are all simple tips you should use every day to protect your TikTok account. Just as important, you should install cyber security controls. These are software programs that protect your devices and your accounts in different ways.

CYBER
SECURITY
CONTROLS

Encryption

Encryption is the process of coding data or messages so that unauthorised users cannot understand them. Encryption allows data to be stored safely on a device or transmitted securely.

On a Windows device, you can encrypt your data by using a Windows tool called BitLocker. BitLocker is included in Windows systems since 2007. When you have enabled **BitLocker**[1], it locks the startup process until you provide a password. This prevents unauthorised users accessing your data.

For Mac devices, **FileVault**[2] is a MacOS encryption app that's equivalent of BitLocker on Windows. FileVault also prevents unauthorised access to your Mac by encrypting the startup disk and requiring a password to unlock it.

If you are transmitting or receiving data, you can secure it by using a **Virtual Private Network (VPN)**, which is encrypted.

Virtual Private Network (VPN)

A Virtual Private Network (VPN) is a service that protects your internet connection and privacy online. VPNs create an encrypted tunnel for your data, protecting your online identity by hiding your IP address, and allow you to use public Wi-Fi hotspots safely.

NordVPN[3] provides a secure, high-quality VPN.

[1] https://learn.microsoft.com/en-us/windows/security/operating-system-security/data-protection/bitlocker/
[2] https://support.apple.com/en-gb/guide/mac-help/mh11785/mac
[3] https://go.nordvpn.net/aff_c?offer_id=15&aff_id=104853&url_id=902

Endpoint software

Endpoint Protection software protects your phone, laptop, tablet, or other device against cyber attacks and data loss. The software examines files and processes on your devices. It checks them for suspicious or malicious activity. If it finds a problem, the software will alert you so you can act.

If you don't have endpoint protection software installed, you are vulnerable to attacks. You could find you are unable to use your TikTok account or any other programs. These attacks come via email, malware, or any of the other vulnerabilities we described earlier in the guide.

With the software in place, you can be confident that you have reduced the risk. Your devices will continue to operate correctly, and you can carry on with all your normal TikTok activities.

One of the best endpoint protection software programs is **ZoneAlarm** [1].

Firewalls

A **firewall** is a software program that protects your devices from bad content. It blocks bad content that tries to reach your devices from the Internet or your Wi-Fi network.

Firewalls operate a set of rules. They use the rules to decide if content is okay or dangerous. Depending on the decision, the firewall blocks the content or allows it through to your device.
ZoneAlarm [1] also offer a very effective firewall program.

[1] https://www.zonealarm.com/?AFFILIATE=221855

Anti-malware software

Anti-malware programs detect malicious software that can damage your data or your device. You probably won't be aware of malware until it's too late. Hackers can install malware code without detection and hiding it inside a device.

Once it's installed, malware can disrupt the operation of your device or allow hackers to gather sensitive information. If malware gains access to other areas of your device, it will cause further problems.

ZoneAlarm[1] provides an effective anti-malware software program.

Anti-virus software

Anti-virus software searches for viruses and other malicious software. If it detects any viruses, it removes them before they get into your device and cause problems. Anti-virus software scans files coming into your computer. It compares them with a huge database of existing viruses. If the software detects a virus, it isolates it and removes it from your device. You can also use the software to scan your device. The scan searches for any viruses that avoided an earlier scan.

If you don't have anti-virus software, viruses can cause your device to crash. The viruses also allow hackers to spy on you or monitor your personal information. Then they could steal your identity or your data.

ZoneAlarm[1] provides strong anti-virus protection.

[1] https://www.zonealarm.com/?AFFILIATE=221855

Choose reliable software

Those software programs play a vital role in protecting your TikTok account, your devices and personal information. It's important to choose software from a reputable supplier. The best suppliers provide advice on using their products. They also keep the software up to date so that it protects you against the latest threat. If you choose a reputable supplier, there is no risk of threats or damage caused by the software.

Don't ignore cyber security

Cyber security is essential for all TikTokers. But what kind of security do you need if you're an influencer or content creator?

DO TIKTOK INFLUENCERS AND CONTENT CREATORS NEED TO CARE MORE ABOUT ONLINE SECURITY THREATS?

TikTok everyday users enjoy watching the latest videos and sharing content too. However, many everyday users want to do more with their TikTok account - get more views, more likes, or even more money. That's when they make the move to become influencers or content creators.

There are certainly great role models to follow — like Khaby Lame with more than 160 million followers or Charlie D'Amelio who has more than 150 million followers on TikTok.

Khaby Lame

Khaby Lame is the most-followed person on TikTok. He hasn't always been famous. He worked in a factory but lost his job during the pandemic in 2020.

Then he turned to social media. He made videos of his amusing responses to complicated life hacks. Making videos wasn't new. Khaby started on YouTube as a child, but his only followers were his family. Then he discovered TikTok. He thinks it's the perfect platform for his comic videos. With more than 160 million followers, he must be right!

Charlie D'Amelio

Charlie D'Amelio is an American social media sensation. She started posting TikTok dancing videos when she was 15 and is now one of the most popular teenagers in the world.

Charlie was a contestant on the US show Dancing with the Stars. She won the competition in November 2023. She was also one of the first TikTokers to feature in a major advertising campaign.

Successful TikTokers know how to stay safe online

The big risks for influencers and content creators

- Losing content.

- Impersonation by criminals.

- Reputation damage.

- Loss of income and followers.

Khaby and Charlie have made the big time. But you don't have to capture those massive numbers to enjoy personal satisfaction and success. We can't teach you how to be a star on TikTok. We can tell you how important online security will be when you start creating content for a larger audience outside of your small circle of followers.

influencers and content creators should follow all the guidelines and tips on basic online security. They also need to take extra measures because they will face risks that don't really affect TikTok everyday users.

Losing content

Content is the most valuable commodity for influencers and content creators. Protect it at all costs! Hackers use various phishing techniques to access the account. Once they've gained access, hackers copy content and transfer it to another site.

Losing content means someone else could use it or change it to discredit the original creator. Content creators must protect content in ways we'll explain later and store it in a secure location.

Store data and content safely

Hackers accessed data of **100,000 influencers** following a breach at a social media marketing firm. The information included influencers' social media links, email addresses, names, phone numbers, and home addresses.

The hackers also stole the data of **250,000 social media users**. They exposed the information on a hacking forum that placed the users at risk of scams.

How 'Steve's' content found a new owner.

Steve is a content creator. He posts popular fitness videos on TikTok. He noticed the demand for fitness videos during the pandemic. Steve has grown his followers ever since. The content has changed since the pandemic. Back then, basic fitness was important for people who were house bound during lockdown. Now Steve makes videos for fitness enthusiasts. He sets his new followers more demanding exercises.

The business was going well. Numbers for his advanced videos grew steadily. But suddenly there was a sharp drop in the number of views of his basic fitness videos. Steve did some Internet research. He found the same videos on offer under the name of a different influencer.Even worse, the criminals had used photo manipulation to change Steve's face. If there's money involved, criminals will go to great lengths!

This is a real story about an individual. We've changed the name and some of the details to protect their identity.

Impersonation by criminals

Social media impersonation is big business

- The FBI estimates that impersonation attacks have caused global losses over **$5.3 billion**.

- Impersonation attacks targeting brands increased by more than **300%** in 2022.

That was one example of content theft and impersonation. It could be worse. Criminals use social media platforms like TikTok to impersonate an influencer who is trusted by followers.

The scammers set up a fake account and profile that imitates the real influencer. They make the fake site look as convincing as possible. Hackers then use that account to scam people who click on a fake link.

The problem is that followers who are scammed blame the influencer or content creator. That causes damage to the influencer's reputation.

'Jenny's' impersonator stole her business.

Jenny makes fashion videos on TikTok. She works with many well-known brands. As part of her videos, she adds links to the clothes she shows. That makes it easy for followers to buy them. Manufacturers recognised the value of working with Jenny to promote their lines. Her business grew. But things hit the rocks when she found out that criminals had used her name and profile to set up a fake account.

The small change they made was to rename the site Janny. That was enough to fool most TikTokers. The problem was that the links on the Janny videos took users to a fake site. TikTokers paid their money but never saw any clothes. When people realised the scam, the word got back to Jenny and the brands. They were not pleasant messages. It took Jenny months to rebuild trust with her followers and the brands.

This is a real story about an individual. We've changed the name and some of the details to protect their identity.

Reputation damage

For influencers, reputation is everything. Followers trust the influencer. If someone betrays that trust, the influencer's reputation goes down a notch. Followers soon disappear.

The damage occurs for different reasons. The influencer might post something inappropriate. It could be a comment that upsets some of their followers. More likely, someone else posts comments on the account that cause serious damage to the influencer's reputation.

Reputation is too valuable to lose

- **69%** of consumers trust the product recommendations they get from influencers.

- **83%** of consumers would buy products sold by companies with top reputations.

- Only **9%** want to buy from companies with poor reputations.

'Eric' loses goodwill when hackers spread damaging comments.

Eric is a joker. He likes to make fun of celebrities on TikTok in a friendly kind of way. Most celebrities like the publicity. They don't mind the attention from Eric and his followers. In fact, they feel left out if they haven't been 'Eric-ed'. But all that goodwill vanished when damaging comments started to appear on the TikToks.

A hacker had been gathering 'dirt' on some famous people. He hacked Eric's account and started posting rumors and bad stories. Papers picked up the stories. The whole situation spiraled out of control. The celebrity threatened to sue Eric. It was no surprise when Eric lost the trust of other celebrities.

This is a real story about an individual. We've changed the name and some of the details to protect their identity.

Loss of income and followers

To summarise, those risks have a common thread. If things go wrong, influencers and content creators lose followers and income.

The risks are high if hackers steal content and re-use it on a fake site because the content creator loses revenue. If a criminal uses stolen content or impersonates an influencer to scam people, the influencer suffers reputation damage and loses income.

When content fees and revenue from sponsorship are a content creator's main income source, stolen content or reputation damage will cause financial hardship. Influencers and content creators must make sure they're aware of the risks. They must use the strongest online security controls to protect their businesses.

Protect the online presence

- Keep information and content safe.
- Look out for impersonators.
- Protect the reputation.

'Helen's' acting career goes from comedy to tragedy

Helen is a young British actress. She used TikTok to earn money after graduating from drama school. She posted comedy sketches that appealed to younger viewers. Her followers eventually topped the million mark.

Everything was going well until hackers used malware to take control of her account. Helen was no longer able to post her videos on TikTok. As a result, she lost her followers and her income. To make things worse, she faced a ransom demand of thousands of dollars to free the account.

This is a real story about an individual. We've changed the name and some of the details to protect their identity.

Brand ambassadors need even stronger security

Many successful influencers become **brand ambassadors**. Brands want to associate with well- known influencers because they know influencers have followers who trust them. That trust can spread to the brand too.

Top brand ambassadors work with major household names. However, there are plenty of smaller brands looking for successful partners.

When influencers act as brand ambassadors, they will be held to a much higher standard of trust and security. The brand owners need to know that the influencer can protect their confidential information, intellectual property, and reputation.

That's a big responsibility. If anything goes wrong, it could jeopardize the influencer's relationship with the brand, or worse.Suppose the influencer exposes confidential brand information or attracts damaging comments. The betrayal of trust could lead to claims for damages or potential lawsuits. That makes cyber security an absolute priority for brand ambassadors.

Brands need influencers

- Consumers trust TikTok videos because they are created by other consumers.

- **37%** of consumers trust social media influencers over brands.

- Generation Z and Millennials are two times more likely than Boomers to trust influencers.

- **84%** of Generation Z have purchased products in direct response to social media content.

'Jamie' exposes a brand to damaging comments

Jamie built a great partnership with a well-known brand of cleaning products. Her content focused on how easy it was to use the products. She also answered questions from followers. One question needed a detailed answer about the product's 'ingredients'. Jamie posted a complete list.

While that was an honest answer, the brand owners warned her about overstepping the mark. They thought she had given away too much information. Things got worse when negative comments appeared. They attacked the use of certain materials in the product. Jamie had no security controls in place to manage comments. The negative posts had not come from Jamie's followers. They came from environmental activists. When the brand lost share, Jamie lost the contract.

This is a real story about an individual. We've changed the name and some of the details to protect their identity.

Strengthen online security

TikTokers who grow their channel as an influencer or content creator will find that online security becomes a bigger part of their lives. As we showed in the stories, weak online security leads to very damaging consequences.

HOW DIFFERENT IS ONLINE SECURITY FOR INFLUENCERS AND CONTENT CREATORS?

> I think that the more influence people begin to have, the more responsibility they must have regarding the content that is being put out.

Influencers have a special responsibility to provide useful information and content to their followers. They are also responsible for ensuring that the way they interact with followers on TikTok is safe and secure.

That makes online security key to achieving success on TikTok. Influencers and content creators need to protect their content, identity, reputation, and the safety of their followers. Influencers don't just have to protect their own personal information and data. They must protect the data and reputations of their brand partners.

TikTok online security considerations for influencers and content creators

Influencers and content creators should follow all the guidelines that protect TikTok everyday users. Firewalls, anti-virus, encryption, and anti-malware software are essential basics. Now, online security measures must go above and beyond the basics. These are some of the important measures to take.

Online security for influencers and content creators

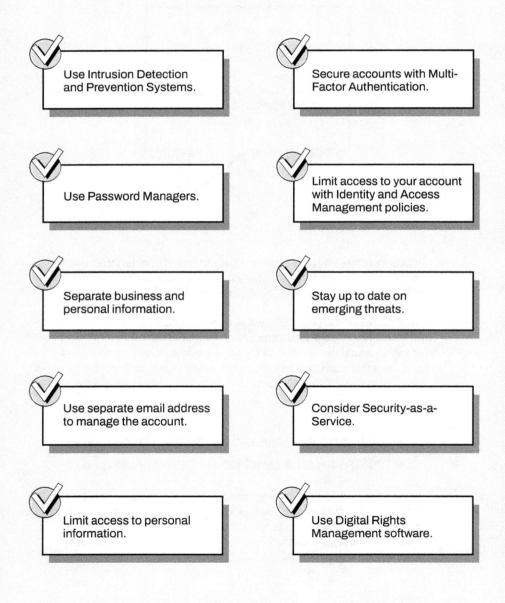

Use Intrusion Detection and Prevention Systems.

Secure accounts with Multi-Factor Authentication.

Use Password Managers.

Limit access to your account with Identity and Access Management policies.

Separate business and personal information.

Stay up to date on emerging threats.

Use separate email address to manage the account.

Consider Security-as-a-Service.

Limit access to personal information.

Use Digital Rights Management software.

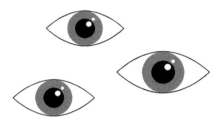

Intrusion Detection and Prevention Systems

An **Intrusion Detection and Prevention System (IDS/ IPS)** is an application that monitors devices and networks, detects suspicious activities, and generates alerts when they are detected. IDS/IPS products provide an early warning of possible attacks. If they detect a threat, they automatically take action to stop it reaching the devices.

Password Managers

Password Managers like **NordPass**[1] provide a more secure method of creating, storing, and using passwords. They store the passwords in a secure software program. Users access the software by entering a single 'master' password. Password Managers are even better when they're used with a password generator. The generator creates very strong complex passwords that could be difficult to remember.

[1]https://go.nordpass.io/aff_c?offer_id=488&aff_id=104853&url_id=9356

Separate business and personal information

It's essential to use different contact details to separate business from personal communications. If there is a cyber attack, the damage will affect the business rather than the content creator's personal life.

Use a separate mobile number to manage your TikTok account. Having a different mobile number to manage the TikTok account will mitigate the risk of cloning and unauthorised access to the account. If the mobile number gets cloned, only the TikToker and TikTok know that that particular number is being used to manage the account.

Limit access to personal information

Make sure the TikTok profile includes only essential personal information. Keep personal and financial information out of emails or posts. Remember hackers can build targets from small snippets of information. Make it difficult for them.

Secure accounts with Multi-Factor Authentication

A single password might be strong. When a hacker discovers it, they're free to attack the TikTok account. Two-factor verification is the absolute minimum. After logging-in with a password, enter a second piece of information. This could be a code or One-Time Password that TikTok sends.

Using biometric login tools like facial or fingerprint recognition will make access control even stronger.

Limit access to the account

Many content creators work with a team of collaborators. That makes it essential to limit and monitor the number of people who can access the creator's accounts. By providing collaborators with unique login credentials, it's easier to monitor access. Creators should also check that collaborators have security levels that are as strong as their own. If collaborators hold creators' information on their sites, they must have security to prevent attacks.

Identity and Access Management (IAM) policies control access by collaborators. An IAM system gives different levels of access to users. Some collaborators may not need access to highly confidential information. The system monitors access so creators always know who is entering their system.

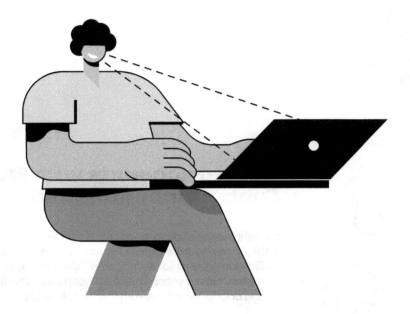

Stay up to date on emerging threats

New cyber security threats emerge every day. It's important for content creators to stay up to date on emerging threats so they are aware and can take measures to reduce new risks.

Reputable suppliers of anti-virus or anti-malware software like **ZoneAlarm**[1] automatically update their programs with data on the latest threats. To keep device security up to date, it's essential to apply any updates that manufacturers notify.

Consider Security-as-a-Service

Security-as-a-Service (SaaS) is a service that takes care of cyber security monitoring and management. The security company experts monitor their customers' devices, accounts, and social media activities. If they identify issues, they deal with them.

[1] https://www.zonealarm.com/?AFFILIATE=221855

Use Digital Rights Management (DRM) software

Digital Rights Management (DRM) software protects content. It ensures people use content correctly and protects revenue. DRM software tracks and identifies who is accessing your content and how they are using it. This will help content creators prevent unauthorised users from copying, downloading, or sharing content with other people without permission. DRM software makes sure the creator receives all the revenue that's due.

Online security matters

Those online security solutions will become increasingly important as the TikTok business grows. By implementing strong security from the start, influencers and content creators can build their TikTok business with confidence.

There is one more factor to consider – regulations. Ignoring them could harm the TikTok business.

BE AWARE OF REGULATIONS

TikTok influencers and content creators must be aware of the regulations that apply to their businesses. Regulations vary by country and by the age of the audience.

In general, there are regulations that cover:

- Privacy.
- Data protection.
- Harmful content.
- Misleading content.
- Content for younger viewers.

There are also guidelines to cover products such as:

- Health-related products.
- Medications.
- Harmful products.

Regulations vary by country

It's important to understand regulations in the local territory. International influencers should check the regulations in other countries where their videos appear.

Who regulates influencers

USA
The Federal Trade Commission (FTC).

UK
The Competition and Markets Authority (CMA).

Australia
The Australian Competition and Consumer Commission.

Compliance is essential

It's not necessary to go into regulations in detail, but it is essential to understand the spirit of the regulations.

Influencers must be honest and clear upfront about any sponsorship or financial benefit they get from the video. Above all, they must respect their followers. Influencers who mislead their audience could face unforeseen problems.

Regulators are paying more and more attention to social media influencers. Content creators should be aware that someone, somewhere could be checking their videos. If they consider the videos are not complying with regulations, there are consequences:

- Fines for breaking regulations.

- Penalties that could damage reputation and the trust of followers.

- Loss of revenue and followers.

Regulation may seem like an inconvenience, but it could influence the future of the business.

THE END GAME

As we've seen throughout this book, TikTok everyday users, influencers and content creators face many challenges to their security and safety. The bad actors who want to steal content, money or personal information can make TikTok an unpleasant experience for their victims.

However, by following the guidelines and using reputable online security methodologies, all TikTokers can prevent damaging, distressing attacks.

We hope you've enjoyed this guide and use the recommendations we suggest. The social media landscape is dangerous, so it's essential you protect yourself, your account, and your followers with strong online security.

With the right security choices, you will make your TikTok experience safer and more enjoyable for everyone.

If you have a moment, we would greatly appreciate it if you could leave a review on the platform you've chosen to purchase this book. Your feedback not only helps us improve our content but also helps other readers discover our work.

ABOUT THE COMPANY

WeCyberYou! is an Australian cyber security media platform that was created to encourage businesses and individuals around the world to be safe online, helping them to protect their interests and brands from malicious code and actors via educational videos, articles, podcasts, books, training, social media content, and services.

Its founder, CEO and the author of this book, Edson Agostinho, is a cyber security enthusiast who has been working in the industry for over 25 years. Edson has been assisting businesses in consulting, finance, education, telecommunications, and information technology as well as state and federal government authorities in Australia to protect their environments and reputations from internal and external cyber threats.

Edson currently holds a degree in system analysis and a postgraduate degree in cyber security in addition to the Certified Information Systems Security Professional (CISSP) and Certified Information Security Manager (CISM) certifications, and other cloud security international credentials.

DEDICATION

This book is dedicated to the unwavering support of my mother, Eneida Agostinho, and my beloved sons, Gabriel, and Paul. To my dear friends whose encouragement has been a constant source of inspiration throughout this journey, I offer my heartfelt gratitude.

Moreover, I extend this dedication to the vibrant community of TikTok influencers and content creators worldwide. Your relentless dedication to producing quality content, uplifting others, and amplifying important messages deserves recognition. You are the driving force behind innovation and empowerment in the digital era and, for that, I salute you.

GLOSSARY

2-step verification

a secure form of login that requires you to enter a password followed by a second piece of information.

Anti-Malware software

a product that protects your devices against malicious software that can damage them or give criminals access to confidential information.

Anti-Virus software

a product that protects your devices against viruses that could affect their performance or security.

Bad actors

individuals or organisations who attack TikTok accounts.

Biometric credentials

a secure form of login that uses biometric features that cannot be copied or stolen.

Brand ambassadors

influencers who work on behalf of brands to raise interest in the brand among their followers.

Compromised login

login that no longer works because hackers have blocked access to an account.

Content creators

TikTokers who produce videos.

Cyber crime

crime committed online.

Cyber criminals

criminals who carry out attacks online.

Cyber security software

products that protect your devices against attacks by criminals or hackers.

Cyber bullies

individuals who use TikTok to attack other TikTok users, causing them distress.

Dark web

a network of websites operated by criminals, used to sell stolen credentials, or transmit crime-related communications.

Digital Rights Management

products that help content creators track and monitor the use of their content.

Encryption

the process of coding data or messages so that they cannot be understood by unauthorised users. Encryption allows data or message to be stored safely on a device or transmitted securely.

Endpoint protection software

products that protect your devices against attack.

Facial recognition

a secure form of login that uses your facial features as a credential.

Fingerprint recognition

a secure form of login that uses your fingerprint as a credential.

Firewall

a product that prevents damaging content from accessing your devices from the Internet.

Hackers

individuals who attack social media accounts.

Identity and Access Management

a product that helps you monitor and control who can access your account.

Identity theft

an online attack that steals your personal information.

Impersonation

use of a stolen TikTok identity.

Influencers

TikTokers with specialist knowledge or expertise on a topic.

Intrusion Detection and Prevention Systems

a product that monitors devices, detects suspicious activities, and takes action to stop threats reaching the devices.

Keystrokes

information created when you type into a device. Hackers can capture the information to steal data.

Malware

malicious software that can damage your devices or enable hackers to steal information from a device.

Multi-Factor Authentication

a secure form of login that requires you to enter two or more pieces of information to gain access.

Online security

policies, products, and guidelines to protect you and your devices online.

Operating System

software that enables your devices to function.

Password Generator

an application that creates strong passwords.

Password Manager

an application to store your passwords securely.

Phishing

a form of online attack designed to gather personal or confidential information.

Private profile

a TikTok profile that is only visible to TikTok users you select.

Public profile

a TikTok profile that any TikTok user can view.

Ransomware

software that enables criminals to 'lock' your devices so you cannot use them until you pay a ransom.

Scams

online attacks that trick people into giving away information or money.

Security breach

a loss of personal information or other data following a cyber attack.

Social engineering

a form of online attack where hackers collect information by persuading victims to trust them and take actions that will ultimately harm them.

Software-as-a-Service

a professional service that takes responsibility for managing security on your behalf.

State actors

criminals who operate online with the support of their governments.

Trolls

individuals who cause distress to TikTokers by posting damaging comments or threats.

Virus

a type of software that infects your devices and damages your data or software.

Virtual Private Network (VPN)

a service that protects your internet connection and privacy online. VPNs create an encrypted tunnel for your data, protect your online identity by hiding your IP address, and allow you to use public Wi-Fi hotspots safely.

**RESEARCH
REFERENCES**

Delete Cyberbullying - How to spot and stop cyberbullying on TikTok
https://www.endcyberbullying.
net/blog/how-to-spot-and-stop-
cyberbullying-on-tiktok

Youthopia - Cyberbullying on TikTok is a major issue
https://youthopia.sg/read/
cyberbullying- on-tiktok-is-a-major-
issue/

Norton - 11 social media threats and scams to watch out for
https://uk.norton.com/blog/online-
scams/11-social-media-threats-and-
scams-to-watch-out-for

LinkedIn - Common cyber security risks associated with social media use and how individuals can protect themselves
https://www.linkedin.com/pulse/
common-cyber-security-risks-
associated-social-media-use-how-
px6bf/

Panda Security - Social media threats
https://www.pandasecurity.com/en/
mediacenter/social-media-threats/

BluSkills - Uncompromising security - reassuring protection
https://www.bluskills.co.uk/case-
studies-1/2023/3/1/uncompromising-
security-reassuring-protection

Security Magazine - How to risk security risks for social media influencers
https://www.securitymagazine.
com/articles/97279-how-to-reduce-
security-risks-for-social-media-
influencers

Creator Wizard - 10 privacy and safety tips for influencers
https://www.creatorwizard.com/
post/10-privacy-safety-tips-for-
influencers

Government of Canada - Cyber security for influencers: How social media personalities can protect themselves online
https://www.getcybersafe.gc.ca/en/
blogs/cyber-security-influencers-
how-social-media-personalities-can-
protect-themselves-online

Concentric - The evolving security paradigm for influencers and gamers
https://www.concentric.io/blog/
the-evolving-security-paradigm-for-
influencers-and-gamers

Turingsecure - Cybersecurity for influencers
https://turingsecure.com/blog/
cybersecurity-for-influencers/

Curious Blogger - How to protect digital content (A guide for influencers)
https://curiousblogger.com/protect-
digital-content/

Vitrium Systems - Content protection with DRM controls
https://www.vitrium.com/lp/
drm?utm_term=digital%20rights%20
management%20software&utm_
campaign=Google_UK_No.1_
FY2024+%23test2&utm_

Social Media Today - What the heck is "Social DRM"?
https://www.socialmediatoday.com/
content/what-heck-social-drm

Tateeda - Digital Rights Management software development: functions, features & services
https://tateeda.com/blog/digital-rights-
management-development

Collab Asia - Services
https://www.collabasia.co/services

Cisco Systems - What is Identity Access Management (IAM)
https://www.cisco.com/c/en_uk/products/security/identity-services-engine/what-is-identity-access-management.html

Redhat - What is an Intrusion Detection and Prevention System (IDPS)
https://www.redhat.com/en/topics/security/what-is-an-IDPS

Gartner - Intrusion Prevention Systems
https://www.gartner.com/reviews/market/intrusion-prevention-systems

Kaspersky - Definitions -Firewall
https://www.kaspersky.com/resource-center/definitions/firewall

Millionaire Mentor - TikTok CEO Shou Zi Chew promises to protect user data
https://millmentor.com/tiktok-ceo-shou-zi-chew-promises-protect-user-data/

TikTok – Video Privacy
https://support.tiktok.com/en/account-and-privacy/account-privacy- settings/video-visibility

TikTok – Account privacy settings
https://support.tiktok.com/en/account-and-privacy/account-privacy-settings

TikTok – Making your account public or private
https://support.tiktok.com/en/account-and-privacy/account-privacy-settings/making-your-account-public-or-private

TikTok Newsroom – How we secure personal information and store data
https://newsroom.tiktok.com/en-us/tiktok-facts-how-we-secure-personal-information-and-store-data

Kaspersky – Preemptive safety – TikTok tips
https://www.kaspersky.co.uk/resource-center/preemptive-safety/tiktok-tips

Malwarbytes labs - TikTok is "unacceptable security risk" and should be removed from app stores, says FCC
https://www.malwarebytes.com/blog/news/2022/07/tiktok-is-unacceptable-security-risk-and-should-be-removed-from-app-stores-says-fcc

Norton - Is TikTok safe? 3 TikTok scams to be aware of in 2024
https://us.norton.com/blog/privacy/is-tiktok-safe#:~:text=Generally%20speaking%2C%20TikTok%20is%20as,given%20private%20accounts%20by%20default

TikTok Newsroom – The truth about TikTok
https://newsroom.tiktok.com/en-au/the-truth-about-tiktok

The Standard - Is TikTok safe to use? Concerns raised about harmful content and data privacy
https://www.standard.co.uk/news/tech/tiktok-safety-content-misinformation-data-children-government-china-b1047589.html#:~:text=A%20paper%20by%20cybersecurity%20firm,on%20servers%20in%20mainland%20China.

USA Today - Addicted to TikTok but worried about China and your data? How to use it safely
https://eu.usatoday.com/story/tech/columnist/komando/2023/06/29/tiktok-china-protect-your-data/70351963007/

Statista – Number of global TikTok users
https://www.statista.com/statistics/1327116/number-of-global-tiktok-users/

Wikipedia – List of most followed TikTok accounts
https://en.wikipedia.org/wiki/List_of_most-followed_TikTok_accounts

The Social Shepherd – TikTok statistics
https://thesocialshepherd.com/blog/tiktok-statistics

AAG - The latest 2024 cyber crime statistics
https://aag-it.com/the-latest-cyber-crime-statistics/

WeCyberYou!
Cybering the Future.

🌐 www.wecyberyou.com
🌐 www.wecyberyou.com.br
✉ info@wecyberyou.com

THE ULTIMATE
CYBER SECURITY GUIDE
FOR TIKTOK EVERYDAY USERS, INFLUENCERS AND CONTENT CREATORS

WeCyberYou! is an Australian cyber security media platform that was created to encourage businesses and individuals around the world to be safe online, helping them to protect their interests and brands from malicious code and actors via educational videos, articles, podcasts, books, training, social media content, and services.

Its founder, CEO and the author of this book, Edson Agostinho, is a cyber security enthusiast who has been working in the industry for over 25 years. Edson has been assisting businesses in consulting, finance, education, telecommunications, and information technology as well as state and federal government authorities in Australia to protect their environments and reputations from internal and external cyber threats.

Edson currently holds a degree in system analysis and a postgraduate degree in cyber security in addition to the Certified Information Systems Security Professional (CISSP) and Certified Information Security Manager (CISM) certifications, and other cloud security international credentials.

This unique book is packed full of expert information that will make your TikTok experience safer and more enjoyable.

This book is for you. It's written specially for all TikTok everyday users, influencers and content creators around the world. We'll show you how you can make the most of TikTok - safely! You'll find the knowledge and guidance you need to protect your TikTok account effectively. Think of it as your personal TikTok safety guide.

ISBN 978-1-76365-282-8

www.ingramcontent.com/pod-product-compliance
Lightning Source LLC
Chambersburg PA
CBHW071259050326
40690CB00011B/2469